Amer Taqa, Banan N. Alhussary, Ghada A. Taqa

Extraction of Pure Ketamine Powder and Study their Analgesic Effect as a Gel on Mice Using a Hot – Plate Test

Der GRIN Verlag publiziert seit 1998 wissenschaftliche Arbeiten von Studenten, Hochschullehrern und anderen Akademikern als eBook und gedrucktes Buch. Die Verlagswebsite www.grin.com ist die ideale Plattform zur Veröffentlichung von Hausarbeiten, Abschlussarbeiten, wissenschaftlichen Aufsätzen, Dissertationen und Fachbüchern.

Document Nr. V209705

Amer Taqa, Banan N. Alhussary, Ghada A. Taqa

Extraction of Pure Ketamine Powder and Study their Analgesic Effect as a Gel on Mice Using a Hot – Plate Test

GRIN Verlag

Die Deutsche Bibliothek verzeichnet diese Publikation in der Deutschen Nationalbibliografie;
detaillierte bibliografische Daten sind im Internet über http://dnb.d-nb.de/ abrufbar.

1. Auflage 2012
Copyright © 2012 GRIN Verlag GmbH
http://www.grin.com
Druck und Bindung: Books on Demand GmbH, Norderstedt Germany
ISBN 978-3-656-38280-5

Extraction of Pure Ketamine Powder and Study their Analgesic Effect as a Gel on Mice Using a Hot –Plate Test

Banan N. Alhussary   Ghada A. Taqa   Amer A. Taqa

Department of Dental Basic Sciences, College of Dentistry. University of Mosul, Mosul, Iraq

ABSTRACT

The present study was undertaken to extract ketamine powder from ketamine hydrochloride by precipitate ketamine. After that we examine the purity of this powder by infra-red (FTIR) and ultra-violet(UV) spectroscopy. ketamine gel in different concentrations was prepared (0.5 , 1 , 5 , 10 , 15)% to evaluate the antinociceptive activity. ketamine powder was seen is pure and this show in infra-red and ultra-violet scanner. Ketamine gel at concentrations 0.5, 1, 5, 10,15) % produce antinociceptive in mice (5.6±2.2) (4.4±2.0) (8.2±4.3) (10.6±5.2) (8±2.1) second after 2 min respectively by using a hot plate test in comparison with control(2.4±2). The percentage of maximum possible effect (MPE) increased from (9.9) % in control group to (23.3) (18.3) (34.2) (44.2) (33.3)% respectively according to the concentrations of ketamine gel after 2 min . Purification of ketamine powder from ketamine solution and use as a gel to could be of value relief pain by topical application.

Key words: Purification of ketamine, Ketamine gel , analgesia , Infrared spectroscopy, Hot-plate test .

1

INTRODUCTION

Ketamine, a phencyclidine (PCP) analog, has been used for more than 30 years to produce "dissociative" anesthesia [1]. In this state the patient is awake and can respond to stimuli but has a diminished sense of awareness and an amnesia for events occurring while under the influence of ketamine.Early experience with ketamine revealed that it also produced analgesia that sometimes well outlasted its anesthetic effects. Although the mechanisms of ketamine's analgesic effects remain the subject of debate, and are likely multiple[2-4]. Antagonism at the NMDA-receptor site appears to be central to both its anesthetic and analgesic effects [5,6]. Ketamine is available only I.M , I.V administration but it has been used orally. Ketamine is a dissociative anesthetic that is used to provide sedation and anesthesia in short surgical procedure, Patient may have adverse psychological effect including hallucinations, nightmares, delusion, dissociative reaction and schizophrenic form psychosis [7]. Ketamine is primarily used for the induction and maintenance of general anesthesia, usually in combination with a sedative. Other uses include sedation in intensive care, analgesia (particularly in emergency medicine), and treatment of bronchospasm. It has been shown to be effective in treating depression in patients with bipolar disorder who have not responded to other anti-depressants [8] . Pharmacologically, ketamine is classified as an N-Methyl D-Aspartat (NMDA) receptor antagonist [9]. Ketamine is an antagonist of N-methyl 1-D-Aspartate (NMDA) class of glutamate receptors which is largely responsible for it's anesthetic and behavioral effect effects [7].NMDA inhibition produce catalepsy ,consistent with the effect of ketamine administration. Ketamine also produces profound analgesia which seen to be at least partially mediated by m- opioid receptor , in addition to it's binding to the phenylcyclidine binding site on the NMDA. ketamine is not frequently used for treatment of humans ,because it induces psychedelic episodes in patients ,especially adults, there are an increasing number of reports about patients that have become addicted to ketamine [10]. Among the latest innovations of the pharmaceutical industry is the technology of drug delivery that overcomes the disadvantages of oral administration, these effects include first-pass metabolism and adverse drug side effects [11]. An ultimate route of administration that by pass these events would offer the patient drug delivery through skin has been a promising concept for a long time because skin easy to access , has a large surface area with fast exposure to circulating and lymphatic net works and the rout is noninvasive [12].Therefore, the aim of the present study is to extract ketamine powder

from ketamine hydrochloride solution and prepare ketamine gel with examine the analgesic effects of topical gel application in mice.

MATERIALS AND METHODS

Preparation of ketamine powder
10 ml of (1M) sodium bicarbonate was slowly added to 100 ml of the aqueous ketamine hydrochloride solution 5% (Alhukamma company) under continuous stirring until the pH of solution was close to 11, stirring was continued for one hour, and then the ketamine was precipitated. The solvent was eliminated by filtration and washed several times with distal water and then dried. The powder was studied by infra-red and ultra-violet spectroscopy in order to confirm the structure of the converted product.

Infrared Spectroscopy.
The infrared spectra recorded for prepared ketamine was examined by using Bruker Tensor 27 IR spectrophotometer (Germany) in the region (500-4000 cm) using KBr disc. This measurement was carried out in University of Mosul, College of Education, Iraq.

Electronic Spectra Measurement:
The measurement was carried out by using ethanol as a solvent with (1cm) diameter quartz cell by using Shimatsu-UV-Vis recording UV-1600 spectrophotometer (Japan). This measurement was carried out in Mosul University, Collage of Science, Department of Chemistry.

Preparation of ketamine gel
Ketamine gel was prepared (0.5, 1, 5, 10, 15)gm of ketamine powder in 100ml gel base to give a final concentration of (0.5%, 1%, 5%, 10%, 15%) with continuous mixing using Vortex device to prepare a homogenous gel. Gels were kept in plastic containers and store at room temperature.

Determination of Analgesic Activity of Ketamine Gel by Using a Hot –Plate Test:
Mice were divided into 2 main groups (A,B) each group were subdivided into six group with 5 animals per each group and the test was assessed by the hot plate method [13]. The mice were treated topically with ketamine gel (0, 0.5, 1, 5, 10, 15) % respectively, on the planter area of the for and hind limb.

3

All the animals in group A were tested on hot plate after 1 minute while the mice in group B were tested after two minutes from topically application of the gel to determine the onset of action of gel. The mice were placed on top of hot plate of $55\pm1\ ^0C$. The time between placement and jumping or licking the hind paw was recorded as response latency.

The reaction time was recorded for control mice and for the animals treated with ketamine gel.

The percentage increase in reaction time was calculated thus using the following equation :- [14]

% increase in reaction time (antinociceptive)=$(T_1 . T_0 / 30\text{-}T_0) \times 100$

T_0 = mean time for the control group (second)

T_1 = mean time for the test group (second)

30= cut off time (second)

Statistical analysis

The data were expressed as mean $\pm$ SD , difference between three experimental groups were statistically analyzed by one way analysis of variance (ANOVA) followed by the least significant difference test. The level of significance was at $p < 0.05$. [15]

RESULTS

Figure (1) represents the vibrational response of pure Ketamine when passed via an infrared beam. The spectrum showed band at 3000-3200cm^{-1} which attributed to the N-H stretch from the amide group connected to the cyclohexanone. The spectrum also showing band at 2800-2900 cm-1 which assign to C-H stretch from an alkyl group. At this frequency the alkyl group is generally a non aromatic CH3 or CH2 stretch. The band at 1750 cm^{-1} due to R_2 –C=O stretch which appear very precise and typical stretch for cyclic ketones. In Ketamine the carbonyl is connected to the cyclohexane ring. The band 1600 cm^{-1} assigned to C-N band (Generally expressed in C-NH2 and C-N=O compounds).The band at 1400-1500 cm^{-1} is attributed to C-H bend, this is another vibration mode of the CH2 or CH3 components of Ketamine. This is not the C-H bond from the aromatic carbons. The band at 1450cm^{-1} region is due to C-C stretch. This carbon to carbon stretch is not for the aromatic specie and hence characterizes the bonding involved in the cyclohexane ring. Pure ketamine extract measurement by UV (Ultra-Violet) spectroscopy Figure (2).

Analgesic effect of ketamine gel

The results of assessment of analgesic activity of ketamine gel at (0.5 , 1 , 5 , 10 , 15)% shown that the gel has no significant difference after 1min from application of gel between control(2.4 ± 2.6) second and all treated groups(1.4 ± 1.6)(2.4 ± 2.4)(5.8 ± 5.4)(4.6 ± 4)(6.2 ± 3.1) second respectively at p< 0.05. Table (1), Figure (3).

The result shown that the application of ketamine gel after 2min produce a highly significant difference between groups treated with ketamine gel at concentration (5, 10, 15) % (8.2 ± 4.3) (10.6 ± 5.9) (8 ± 5.5) second respectively in comparison with control(2.4 ± 2) and other treated group Table (2 and 3).

Topical application of ketamine gel at 0.5% and 10% produce analgesic activity in highly significant difference after 2min (5.6 ± 3.5) (10.6 ± 5.9) sec at p <0.05 in comparison with same concentration after 1min from application (1.4 ± 1.6)(4.6 ± 4)sec respectively at p< 0.05 table (2), Figure(4).

Reaction time (antinociceptive) was increased from (10)% in control treated group after 1min to (25, 19.2, 25.8)% according to the concentration of gel (5, 10, 15)% (figure2), while after 2min the percentage of reaction time (antinociceptive) increase to (23.3, 18.3, 34.2, 44.2, 33.3)% respectively according to the increase concentration of ketamine gel(0.5, 1, 5, 10, 15) in comparison to control group (9.9) % (figure(5).

DISCUSSION

ketamine is classified as an NMDA receptor antagonist [9]. Our findings suggest that topical application of an ointment containing KET within the range of concentration from 0. 5% to 15% is an additional approach to attenuate painful stimulated by thermal (Hot-plate) in mice. The analgesic activity depended on concentration of ketamine. This result agreement with previous study that suggested that topical application ketamine demonstrate efficacy in neuropathic and nociceptive pain[16,17]. The analgesic efficiency of ketamine gel it's may be attributed to it's action on NMDA receptors [7].

Many studies identified several glutamate receptors, such as NMDA, amino-3-hydroxy-5-methyl-4-isoxazolepropionic acid and kainate with action on unmyelinated, myelinated, and postganglionic sympathetic axons. It was suggested that these peripherally distributed receptors play a role in the transmission of sensory signals to the central nervous system. [18,19]. Glutamate is the primary excitatory neurotransmitter of central nervous system and is normally released by pain-signaling afferent

neurons as they synapse on central pain pathways in the spinal cord. The persistence release of glutamate, due to peripheral injury or inflammation, leads to the activation of N-methyl-D-aspartate (NMDA) receptors. This process of activation has been shown to play a crucial in mediating the phenomena of pain[20]. This activation can be prevented mitigated by agents that block the effects of glutamate at NMDA receptor [21,22]. This discoveries have promoted attempted to use NMDA receptor antagonist in the treatment of neurogenic and other, often difficult to control, pain state[23]. NMDA receptor antagonism effects analgesia by preventing central sensitization in dorsal horn neurons; in other words, ketamine's actions interfere with pain transmission in the spinal cord [24] .

Another possible explanation of the analgesic activity of ketamine gel, its interacts with sigma and opioid μ receptors [25-27]. Ketamine also inhibits nitric oxide synthase, inhibiting production of nitric oxide, a neurotransmitter involved in pain perception, and hence further contributing to analgesia [28]. another mechanism of analgesic action of ketamine due to blocks voltage-sensitive calcium channels and depresses sodium channels, attenuating hyperalgesia; it alters cholinergic neurotransmission, which is implicated in pain mechanisms; and it acts as a noradrenergic and serotonergic uptake inhibitor, which are involved in descending antinociceptive pathways [29].

CONCLUSION

Although only a small number of topical agents are available for use in peripheral and local conditions, the obtained results demonstrate that the topical ketamine may provide clinicians with anew option in the battle against pain

REFERENCES

1. Orser BA, Pennefather PS, MacDonald JF.Multiple mechanisms of ketamine blockade of Nmethyl-D-aspartate receptors. Anesthesiology. 1997; 86:903-917.
2. Eide PK, Stubhaug A, Breivik H, Oye I. Reply to S.T. Meller: Ketamine: relief from chronic pain through actions at the NMDA receptor. Pain.1997; 291-72:289
3. Meller ST. Ketamine: relief from chronic pain through actions at the NMDA receptor? Pain 1996,.436-68:435.
4. Pekoe GM, Smith DJ. The involvement of opiate and monoaminergic neuronal systems in the analgesic effects of ketamine. Pain1982;12:57–73.
5. Lodge D, Johnson KM. Noncompetitive excitatory amino acid receptor antagonists. Trends Pharmacol Sci 1990;11:81–86.
6. Klepstad P, Maurset A, Moberg ER, Oye I.Evidence of a role for NMDA receptors in pain perception. Eur J Pharmacol 1990;187:513–518.
7. Yagiela AG, Jone FG, Dowed FJ, Johnson BS, Mariotti AJ, Neidle EA. Pharmacology and therapeutics for dentistry. American Dental Association. New-York 6th ed. 2011; P. 78-161.
8. Nancy Diazgranados et al "A Randomized Add-on Trial of an N-methyl-d-aspartate Antagonist in Treatment-Resistant Bipolar Depression". Archives of General Psychiatry 2010; 67 (8): 793–802.
9. Harrison NL,Simomonds MA February . "Quantitative studies on some antagonists of NMDA inslices of rat cerebral cortex " British journal of pharmacology.1985; 84 (2):381.
10. Jansen, KL "Non-medical use of ketamine.". BMJ (Clinical research ed.) 1993;306 (6878): 601–2.
11. Alarcon E,etal,A comparative study of the transdermal penetration of a series of non steroidal anti-inflammatory drugs ,Journal of pharmacentical scinces ,2002,86 (24)503-308.
12. Daniels R, Knie U , Galenics of dermal product , vehicles , properties and drug release . JDDG .2007; 5 : 367-381.
13. Ghosh MN (1984). Fundamental of experimental pharmacology, Scientific book Agency Calcutta, 2nd edition, p. 144-145.
14. Giusti P, Buriani A, Cima L and Lipartiti M. Effect of acute and chronic tramadol on [3H]-5-HT uptake in rat cortical synaptosomes. Bri J Pharmacology. 1997; 122: 302-306.
15. Daniel WW. Biostatics basic concepts and methodology for health sciences. Wiely J Son .INC 9th edition 2010;P:346.
16. Kronenberg,R. Ketamine as an analgesic :parenteral, oral, rectal, subcutaneous, transdermal and intranasal administration; Journal of Pain& Palliative Care Pharmacotherapy.2002; 16(3) 27-35.

17. Gammaitoni A, Gallagher R, Welz-Bosna M. Topical ketamine gel: possible role in treating neuropathic pain. Pain Med. 2000; 1(1): 97-100.
18. Carlton SM, Chung K, Ding Z, Coggeshall RE. Glu- tamate receptors on postganglionic sympathetic axons. Neuroscience 1998;83:601-605.
19. Coggeshall RE, Carlton SM. Ultrastructural analysis of NMDA, AMPA, and kainate receptors on unmyelinated and myelinated axons in the periphery. J Comp Neurol 1998;391:78-86.
20. Dickenson, AH.. A cure for wind up: NMDA receptor antagonist as potential analgesics. Trends Pharmacol Sci. 1990; 11:307-309.
21. Cherry DA, Plummer JL, Gourlay GK, Coates KR, Odgers CL. Ketamine as an adjunct to morphine in the treatment of pain. Pain 1995;62:119- 121.
22. Lauretti GR, Lima IC, Reis MP, Prado WA, Pereira NL. Oral ketamine and transdermal nitroglycerin as analgesic adjuvants to oral morphine therapy for cancer pain management. Anesthesiology 1999;90:1528–1533,
23. Sang CN, Booher S, Gilron I, Parada S, Max MB. Dextromethorphan and memantine in painful diabetic neuropathy and postherpetic neuralgia: efficacy and dose-response trials. Anesthesiology 2002;96:1053–1061.
24. Quibell, Rachel; Prommer, Eric E., Mihalyo, Mary, Twycross, Robert, Wilcock, Andrew. "Ketamine*". Journal of Pain and Symptom Management.2011; 41 (3): 640–649.
25. Rowland, LM . "Subanesthetic ketamine: how it alters physiology and behavior in humans.". Aviation, space, and environmental medicine. PMID 2005; 76 (7 Suppl): 52-8.
26. Hirota K, Sikand KS, Lambert DG . "Interaction of ketamine with mu2 opioid receptors in SH-SY5Y human neuroblastoma cells". Journal of Anesthesia.1999; 13 (2): 107–9.
27. Narita M, Yoshizawa K, Aoki K, Takagi M, Miyatake M, Suzuki T ."A putative sigma1 receptor antagonist NE-100 attenuates the discriminative stimulus effects of ketamine in rats". Addiction Biology. 2001; 6 (4): 373–376.
28. Aroni, F; Iacovidou, N, Dontas, I, Pourzitaki, C, Xanthos, T "Pharmacological aspects and potential new clinical applications of ketamine: reevaluation of an old drug.". Journal of clinical pharmacology.2009 ; 49 (8): 957–64.
29. Meller, ST . "Ketamine: relief from chronic pain through actions at the NMDA receptor?". Pain. 1996; 68 (2-3): 435–6.

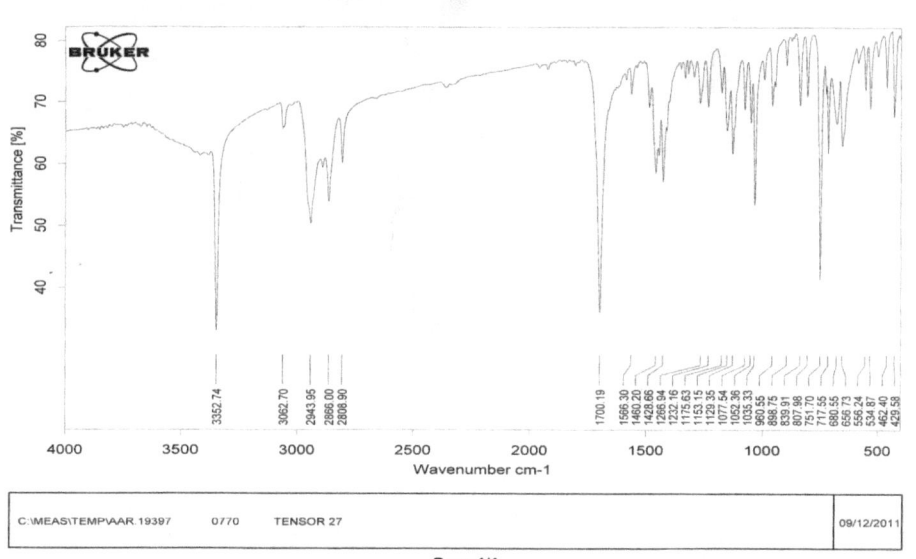

Figure (1): Pure ketamine extract measurement by FTIR (Infra-red) spectroscopy .

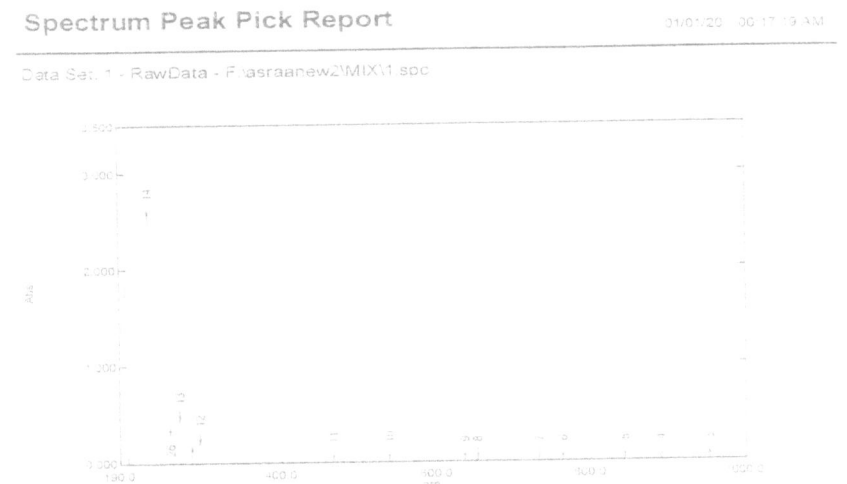

Figure (2): Pure ketamine extract measurement by UV (Ultra-Violet) Spectroscopy.

Table (1): Statistics of antinociceptive effect of ketamine gel after 1
minutes

ANOVA After 1 Minute

	Sum of Squares	df	Mean Square	F–value	p–value
Between Groups	100.400	5	20.080	1.707	0.171
Within Groups	282.400	24	11.767		
Total	382.800	29			

Table (2): Statistics of antinociceptive effect of ketamine gel after 2
Minutes.

ANOVA after 2 Minutes

	Sum of Squares	df	Mean Square	F–value	p–value
Between Groups	219.867	5	43.973	2.503	0.048*
Within Groups	421.600	24	17.567		
Total	641.467	29			

* indicated significant difference at $p \leq 0.05$.

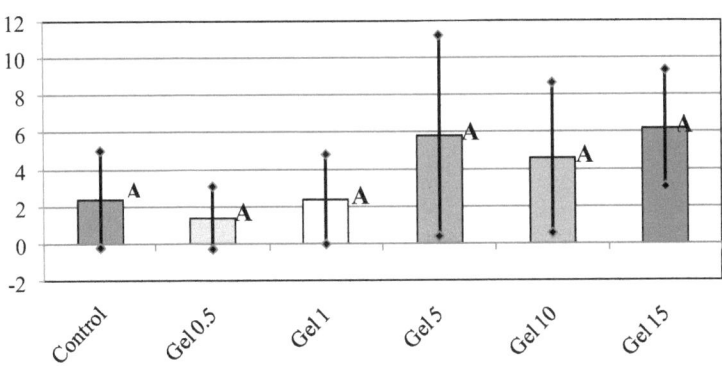

Figure (3): Mean ±S.D of antinociceptive effects of ketamine gel after
1minute in hot-plate test.

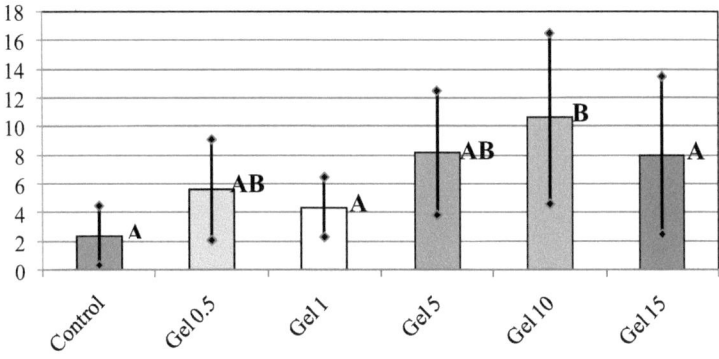

Figure (4): Mean ±S.D of antinociceptive effects of ketamine gel after 2minutes in Hot-plate test.

Table (3): Antinociceptive effect of different concentration of ketamine gel after 1 and 2 min
Hotplate Student's t–test

Concentration	Time	No.	Mean	± SD	t–value	df	p–value
Control	After 1 Minute	5	2.40	2.608	0.000	8	1.000
	After 2 Minutes	5	2.40	2.074			
0.5%	After 1 Minute	5	1.40	1.673	−2.417	8	0.042*
	After 2 Minutes	5	5.60	3.507			
1%	After 1 Minute	5	2.40	2.408	−1.407	8	0.197
	After 2 Minutes	5	4.40	2.074			
5%	After 1 Minute	5	5.80	5.404	−0.775	8	0.460
	After 2 Minutes	5	8.20	4.324			
10%	After 1 Minute	5	4.60	4.037	−1.868	8	0.05*
	After 2 Minutes	5	10.60	5.941			
15%	After 1 Minute	5	6.20	3.114	−0.635	8	0.543
	After 2 Minutes	5	8.00	5.523			

* indicated significant difference at $p \leq 0.05$.

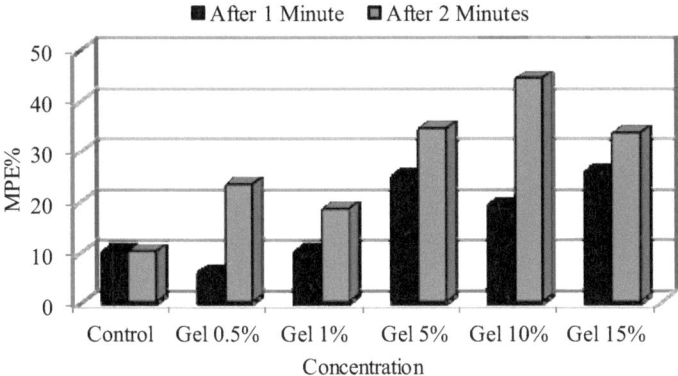

Fig (5): Effect of ketamine gel (0.5 , 1 , 5 , 10 , 15)% on antinociceptive
maximum pain effect (MPE) in hot plate test after 1and 2 min .